Learn or Review

TRIGONOMETRY

Essential Skills

STEP-BY-STEP

Math Tutorials

∞

Chris McMullen, Ph.D.

Learn or Review Trigonometry
Essential Skills

Step-by-Step Math Tutorials
Zishka Publishing

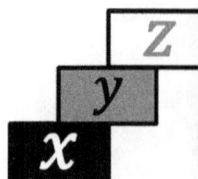

Mathematics > Trigonometry

ISBN-13: 978-1-941691-03-8
ISBN-10: 1-941691-03-X

CONTENTS

INTRODUCTION

This book introduces essential trigonometry skills. Topics include basic trig functions, reference angle, going beyond Quadrant I, inverse trig functions, special angles, radians, and the unit circle.

Several **tips** are included, like how to evaluate trig functions at special angles.

Note that more advanced topics, such as laws, identities, solving equations, and graphs are covered in a subsequent book (and so do not appear in this book).

1 REVIEW OF BASICS

We begin with a review of a few definitions and properties of right triangles that are relevant to trigonometry, including:

∞ acute, right, and obtuse angles
∞ complements and supplements
∞ the sum of the angles of a triangle
∞ complements in a right triangle
∞ the Pythagorean Theorem

An **acute** angle is less than 90°.

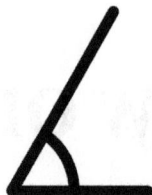

A **right** angle equals 90°; the lines that meet are perpendicular.

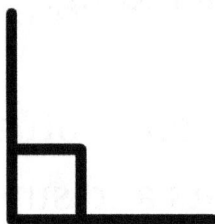

An **obtuse** angle exceeds 90°.

Notation: We often use lowercase Greek letters to represent angles.

α	alpha
β	beta
γ	gamma
θ	theta
φ	phi

Note: Triangles in this book are not drawn to scale. The idea is to think your way through the diagrams, and not guess the angles based on how they appear. This is common in geometry and trig classes, as well as standardized exams.

Complementary angles add up to 90°.

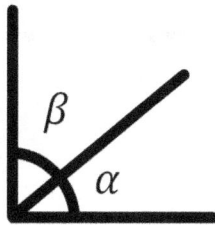

$$\alpha + \beta = 90°$$

Supplementary angles add up to 180°.

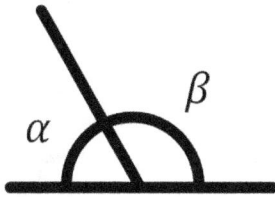

$$\alpha + \beta = 180°$$

The angles of any triangle add up to 180°.

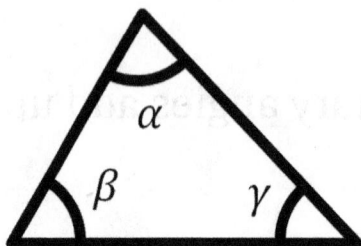

$$\alpha + \beta + \gamma = 180°$$

If you know 2 angles of a triangle, you can use this rule to solve for the third angle.

Example: Find θ in the triangle below.

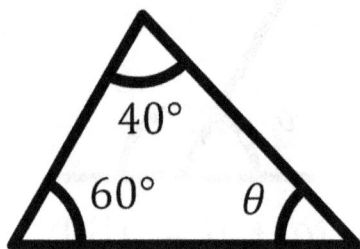

$$40° + 60° + \theta = 180°$$
$$100° + \theta = 180°$$
$$\theta = 80°$$

In a right triangle, the two smaller angles add up to 90°; they are complementary.

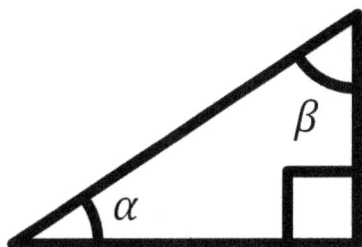

$$\alpha + \beta = 90°$$

Example: Find θ in the triangle below.

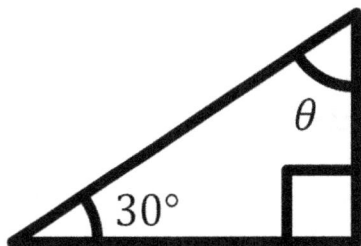

$$30° + \theta = 90°$$
$$\theta = 60°$$

The **Pythagorean Theorem** states that the squares of the legs of a right triangle add up to the square of the hypotenuse.

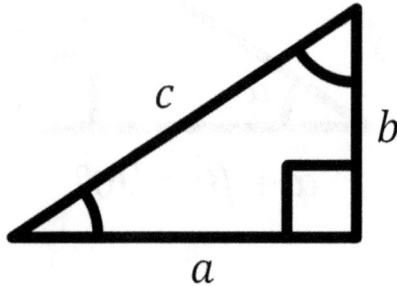

$$a^2 + b^2 = c^2$$

The **hypotenuse** is the longest side of a right triangle; it is the one side that does not touch the 90° angle. The hypotenuse in the above triangle is labeled c.

Example: Solve for c.

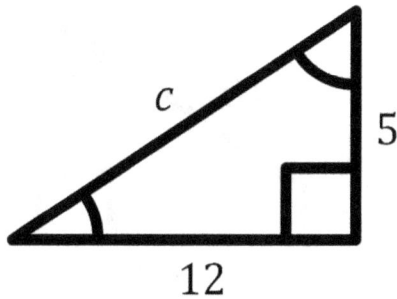

$$5^2 + 12^2 = c^2$$
$$25 + 144 = c^2$$
$$169 = c^2$$
$$\sqrt{169} = c$$
$$13 = c$$

Example: Solve for b.

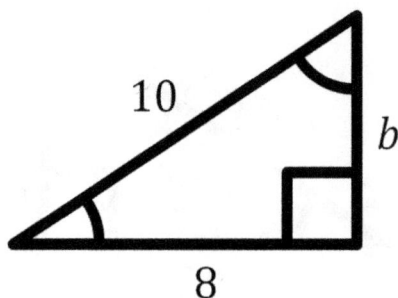

$$8^2 + b^2 = 10^2$$
$$64 + b^2 = 100$$
$$b^2 = 36$$
$$b = \sqrt{36}$$
$$b = 6$$

Practice

Check your answers on page 129.

1. Find the unknown angles:

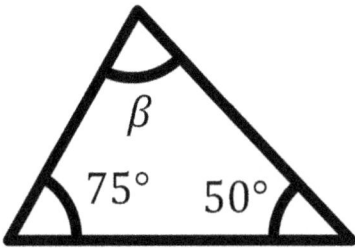

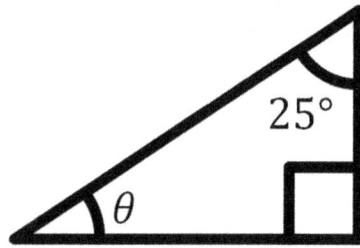

2. Find the unknown sides:

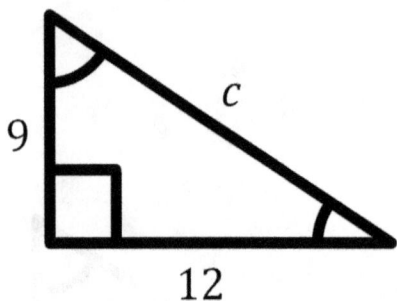

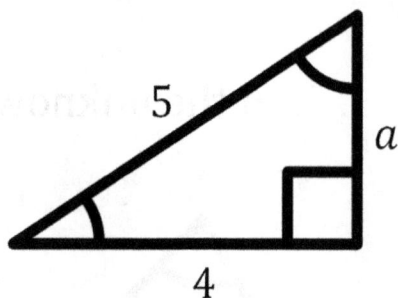

2 BASIC TRIG FUNCTIONS

This chapter introduces you to the basic trig functions, including:

∞ sine, cosine, and tangent
∞ opposite, adjacent, and hypotenuse
∞ trig functions and right triangles
∞ understanding the trig functions

Note: The secant, cosecant, and cotangent functions are introduced in Chapter 6.

Trig functions relate sides of right triangles to one of the angles:

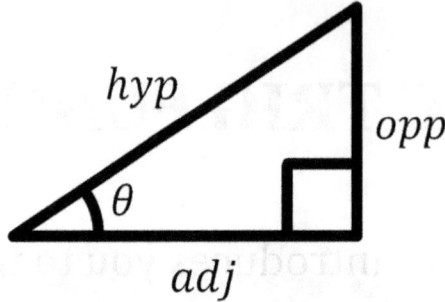

$$\sin\theta = \frac{opp}{hyp}$$

$$\cos\theta = \frac{adj}{hyp}$$

$$\tan\theta = \frac{opp}{adj}$$

∞ adj = adjacent
∞ opp = opposite
∞ hyp = hypotenuse
∞ $\sin\theta$ = sine of theta
∞ $\cos\theta$ = cosine of theta
∞ $\tan\theta$ = tangent of theta

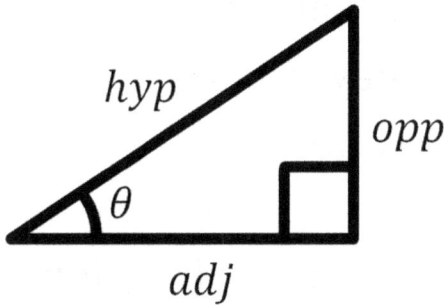

The **hypotenuse** is the longest side; it does not touch the 90° angle.

The **opposite** does not touch the angle θ.

The **adjacent** touches both θ and 90°.

What does it mean?

The sine (sin), cosine (cos), and tangent (tan) functions are **ratios** (they are **fractions**). They tell how one side compares to another.

For example, the sine function is the ratio of the opposite side to the hypotenuse.

The **argument** of the trig function is an angle. For example, in $\sin \theta$, the angle θ is the argument of the sine function.

In the equation $\sin 30° = \frac{1}{2}$, the sine of an angle (30°) results in a fraction $\left(\frac{1}{2}\right)$. The angle has units of degrees (or radians); the result of the function is a fraction with no units.

Finding the Trig Functions

To find **sin *θ*** in a right triangle, divide the **opposite** side by the **hypotenuse**.

To find **cos *θ*** in a right triangle, divide the **adjacent** side by the **hypotenuse**.

To find **tan *θ*** in a right triangle, divide the **opposite** side by the **adjacent**.

The side **opposite** to *θ* does not touch *θ*.

The side **adjacent** to *θ* touches *θ* and 90°.

The **hypotenuse** does not touch 90°.

Example: Find **sin θ, cos θ,** and **tan θ.**

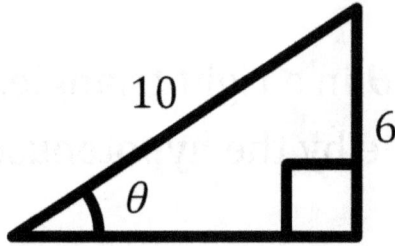

The hypotenuse is 10. The opposite is 6. Use the Pythagorean Theorem to find the adjacent:

$$a^2 + 6^2 = 10^2$$
$$a^2 + 36 = 100$$
$$a^2 = 100 - 36$$
$$a^2 = 64$$
$$a = \sqrt{64}$$
$$a = 8$$

Example continued:

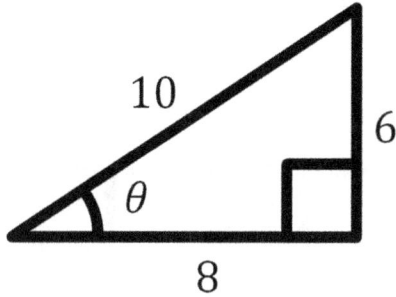

The hypotenuse is 10.

The opposite is 6. The adjacent is 8.

Use the definitions of the trig functions:

$$\sin\theta = \frac{opp}{hyp} = \frac{6}{10} = \frac{3}{5}$$

$$\cos\theta = \frac{adj}{hyp} = \frac{8}{10} = \frac{4}{5}$$

$$\tan\theta = \frac{opp}{adj} = \frac{6}{8} = \frac{3}{4}$$

Example: Find **sin θ**, **cos θ**, and **tan θ**.

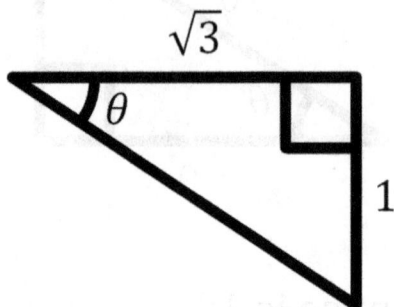

The adjacent is $\sqrt{3}$. The opposite is 1. Use the Pythagorean Theorem to find the hypotenuse:

$$1^2 + \sqrt{3}^2 = c^2$$
$$1 + 3 = c^2$$
$$4 = c^2$$
$$2 = c$$

Example continued:

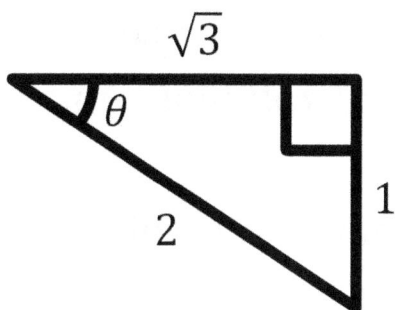

The hypotenuse is 2.

The opposite is 1. The adjacent is $\sqrt{3}$.

Use the definitions of the trig functions:

$$\sin \theta = \frac{opp}{hyp} = \frac{1}{2}$$

$$\cos \theta = \frac{adj}{hyp} = \frac{\sqrt{3}}{2}$$

$$\tan \theta = \frac{opp}{adj} = \frac{1}{\sqrt{3}} = \frac{1}{\sqrt{3}}\frac{\sqrt{3}}{\sqrt{3}} = \frac{\sqrt{3}}{3}$$

Let us briefly review a couple of basic properties of **irrational** numbers.

A squareroot times itself removes the squareroot. For example,

$$\sqrt{3}\sqrt{3} = 3$$

If the denominator is a squareroot, **rationalize** the denominator by multiplying the numerator and denominator by the squareroot. For example,

$$\frac{2}{\sqrt{2}} = \frac{2}{\sqrt{2}}\frac{\sqrt{2}}{\sqrt{2}} = \frac{2\sqrt{2}}{2} = \sqrt{2}$$

Practice

Check your answers on page 130.

1. Find sin θ, cos θ, and tan θ for each triangle.

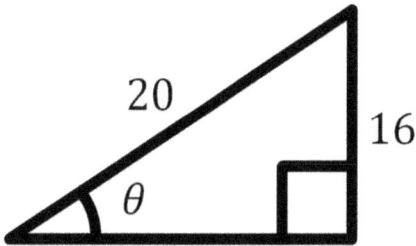

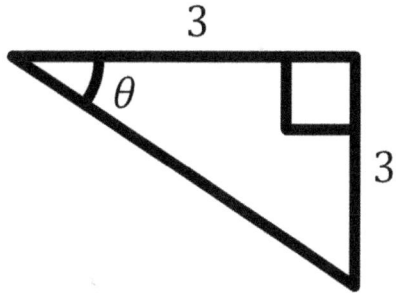

2. Find sin θ, cos θ, and tan θ for each triangle.

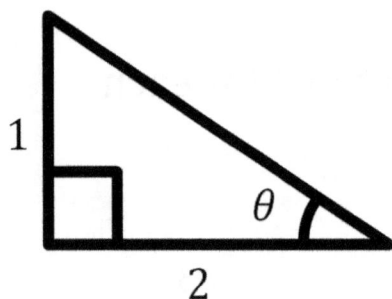

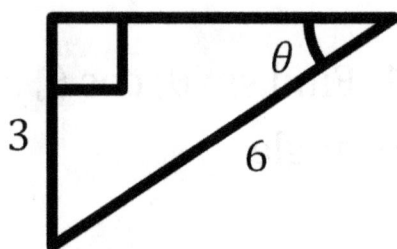

3 SPECIAL ANGLES

Special angles are angles where the trig functions can be evaluated quickly without using a calculator. We will explore the simplest special angles in this chapter, including:

∞ the 45°-45°-90° triangle

∞ the 30°-60°-90° triangle

∞ trig functions for special angles

∞ 0°, 30°, 45°, 60°, and 90°

∞ a memorization tip

Consider the 45° right triangle.

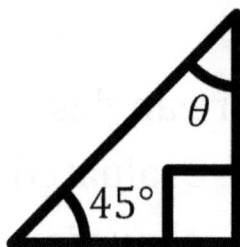

The other angle must also be 45° because the three angles of any triangle must add up to 180°. Since one angle of a right triangle is 90°, the other two angles must be complementary:

$$45° + \theta = 90°$$
$$\theta = 45°$$

The 45°-45°-90° triangle is one-half of a square. (Cut a square along its diagonal.)

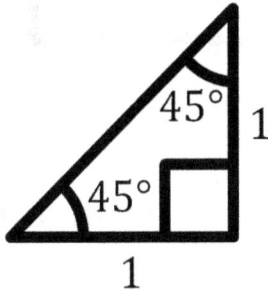

Since two angles are equal to 45°, this is an **isosceles** triangle: two of the sides are equal. If the two legs are each 1 unit long, the hypotenuse can be found from the Pythagorean Theorem:

$$1^2 + 1^2 = c^2$$
$$2 = c^2$$
$$\sqrt{2} = c$$

The hypotenuse is $\sqrt{2}$.

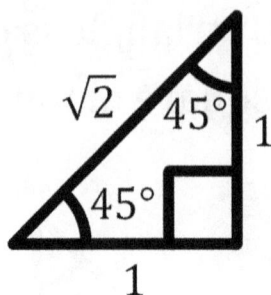

The sine, cosine, and tangent of 45° can be found from the above triangle:

$$\textbf{sin } 45° = \frac{opp}{hyp} = \frac{1}{\sqrt{2}} = \frac{1}{\sqrt{2}}\frac{\sqrt{2}}{\sqrt{2}} = \frac{\sqrt{2}}{2}$$

$$\textbf{cos } 45° = \frac{adj}{hyp} = \frac{1}{\sqrt{2}} = \frac{1}{\sqrt{2}}\frac{\sqrt{2}}{\sqrt{2}} = \frac{\sqrt{2}}{2}$$

$$\textbf{tan } 45° = \frac{opp}{adj} = \frac{1}{1} = 1$$

The sine and cosine functions are equal to each other for 45°. This is because the opposite and adjacent sides are equal for a 45°-45°-90° triangle.

Consider the 30° right triangle.

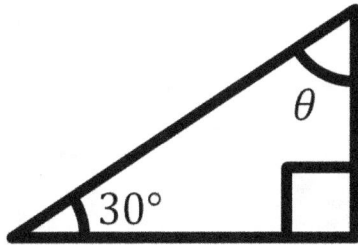

The other angle must be 60° because the three angles of any triangle must add up to 180°. Since one angle of a right triangle is 90°, the other two angles must be complementary:

$$30° + \theta = 90°$$
$$\theta = 60°$$

The 30°-60°-90° triangle is one-half of an equilateral triangle. Therefore, the side opposite to 30° is half the hypotenuse.

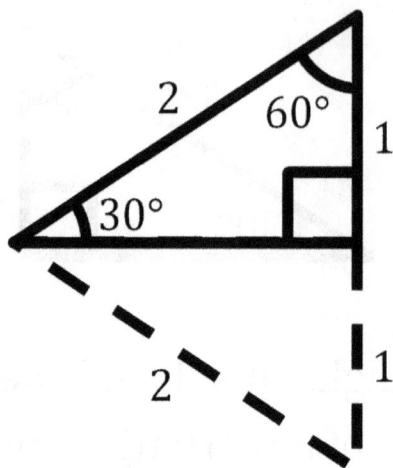

The third side can be found from the Pythagorean Theorem:

$$1^2 + b^2 = 2^2$$
$$1 + b^2 = 4$$
$$b^2 = 3$$
$$b = \sqrt{3}$$

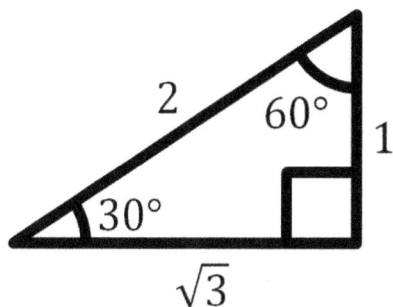

The sine, cosine, and tangent of 30° and 60° can be found from the above triangle:

$$\mathbf{\sin 30°} = \frac{opp}{hyp} = \frac{1}{2}$$

$$\mathbf{\cos 30°} = \frac{adj}{hyp} = \frac{\sqrt{3}}{2}$$

$$\mathbf{\tan 30°} = \frac{opp}{adj} = \frac{1}{\sqrt{3}} = \frac{1}{\sqrt{3}}\frac{\sqrt{3}}{\sqrt{3}} = \frac{\sqrt{3}}{3}$$

$$\mathbf{\sin 60°} = \frac{opp}{hyp} = \frac{\sqrt{3}}{2}$$

$$\mathbf{\cos 60°} = \frac{adj}{hyp} = \frac{1}{2}$$

$$\mathbf{\tan 60°} = \frac{opp}{adj} = \frac{\sqrt{3}}{1} = \sqrt{3}$$

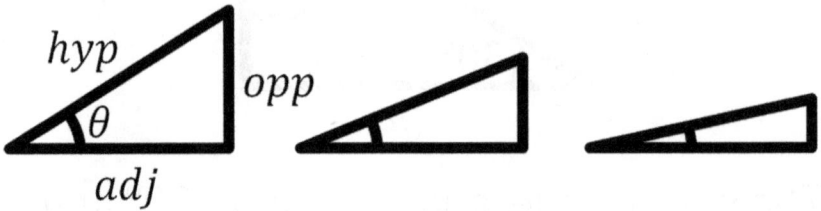

As θ becomes smaller and smaller, the opposite side shrinks to zero while the adjacent side grows closer and closer to the size of the hypotenuse. Therefore,

$$\mathbf{\sin 0°} = \frac{opp}{hyp} = \frac{0}{hyp} = 0$$

$$\mathbf{\cos 0°} = \frac{adj}{hyp} = \frac{hyp}{hyp} = 1$$

$$\mathbf{\tan 0°} = \frac{opp}{adj} = \frac{0}{hyp} = 0$$

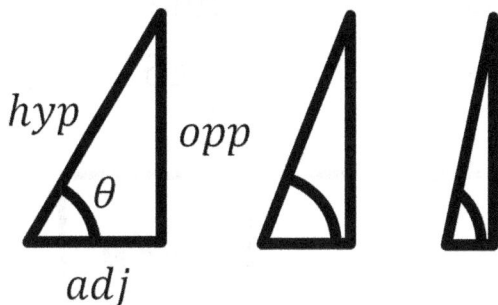

As θ gets closer to 90°, the adjacent side shrinks to zero while the opposite side grows closer and closer to the size of the hypotenuse. Therefore,

$$\sin 90° = \frac{opp}{hyp} = \frac{hyp}{hyp} = 1$$

$$\cos 90° = \frac{adj}{hyp} = \frac{0}{hyp} = 0$$

$$\tan 90° = \frac{opp}{adj} = \frac{hyp}{0} = \text{undefined}$$

The following chart tabulates the basic trig functions at 0°, 30°, 45°, 60°, and 90°.

θ	$\sin\theta$	$\cos\theta$	$\tan\theta$
0°	0	1	0
30°	$\dfrac{1}{2}$	$\dfrac{\sqrt{3}}{2}$	$\dfrac{\sqrt{3}}{3}$
45°	$\dfrac{\sqrt{2}}{2}$	$\dfrac{\sqrt{2}}{2}$	1
60°	$\dfrac{\sqrt{3}}{2}$	$\dfrac{1}{2}$	$\sqrt{3}$
90°	1	0	undef.

Memory tip: There is a simple trick for working out the sine, cosine, and tangent of the special angles.

Follow these steps:

∞ Write 0°, 30°, 45°, 60°, and 90°.
∞ Write the numbers 0 thru 4.
∞ Squareroot each number.
∞ Divide by 2.
∞ These equal sin θ.
∞ Write the numbers backwards.
∞ These equal cos θ.
∞ Divide sin θ by cos θ.
∞ These equal tan θ.

Compare the steps listed above with the table on the following page.

The following tables applies the trick for determining the trig functions at 0°, 30°, 45°, 60°, and 90°.

θ	0°	30°	45°	60°	90°
#'s	0	1	2	3	4
$\sqrt{}$	0	1	$\sqrt{2}$	$\sqrt{3}$	2
$\sin \theta$	0	$\dfrac{1}{2}$	$\dfrac{\sqrt{2}}{2}$	$\dfrac{\sqrt{3}}{2}$	1
$\cos \theta$	1	$\dfrac{\sqrt{3}}{2}$	$\dfrac{\sqrt{2}}{2}$	$\dfrac{1}{2}$	0
$\tan \theta$	0	$\dfrac{\sqrt{3}}{3}$	1	$\sqrt{3}$	und.

Practice

Check your answers on page 131.

1. $\sin 60° = ?$ 2. $\cos 45° = ?$

3. $\tan 45° = ?$ 4. $\sin 0° = ?$

5. $\cos 30° = ?$ 6. $\tan 30° = ?$

7. $\cos 90° = ?$ 8. $\sin 30° = ?$

9. $\tan 60° = ?$

10. $\cos 60° = ?$

11. $\sin 45° = ?$

12. $\tan 0° = ?$

13. $\cos 0° = ?$

14. $\tan 90° = ?$

15. $\sin 90° = ?$

16. $\sin 60° = ?$

4 REFERENCE ANGLE

When working with angles outside of Quadrant I, the reference angle is handy. This chapter describes:

∞ the (x, y) coordinate system
∞ the four Quadrants
∞ how angles are measured
∞ how to find the reference angle

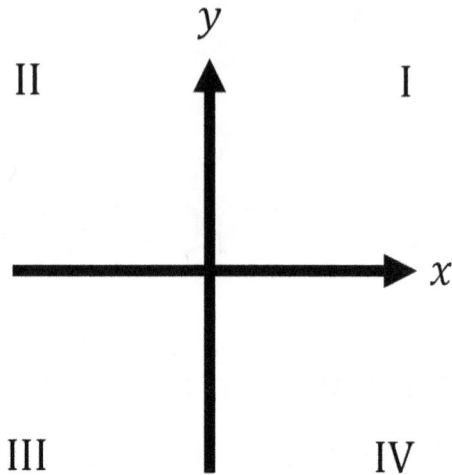

The x- and y-axes define four **Quadrants**, labeled I thru IV in counterclockwise order (as illustrated above):

∞ Quadrant I: x and y are both +
∞ Quadrant II: x is −, y is +
∞ Quadrant III: x and y are both −
∞ Quadrant IV: x is +, y is −

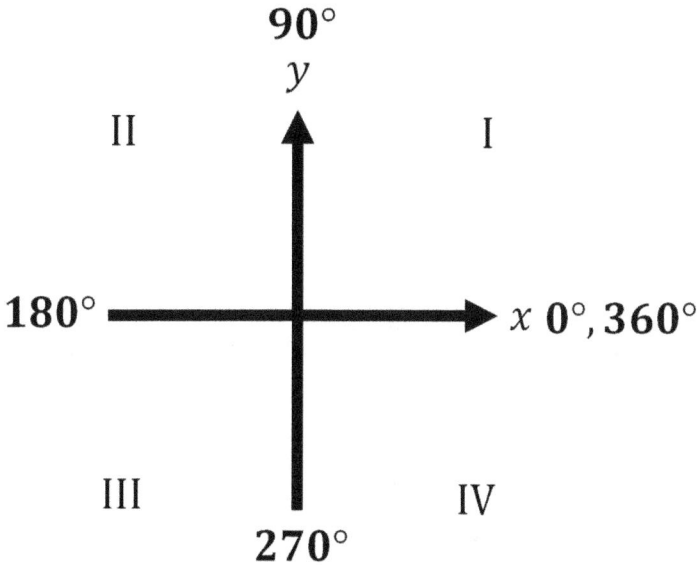

Angles are measured **counterclockwise** from the $+x$-axis, such that:

∞ $+x$ corresponds to 0°

∞ $+y$ corresponds to 90°

∞ $-x$ corresponds to 180°

∞ $-y$ corresponds to 270°

∞ 360° is aligned with 0°

$$90°$$
$$y$$

II   I

$$180° \longleftrightarrow x \ 0°, 360°$$

III   IV

$$270°$$

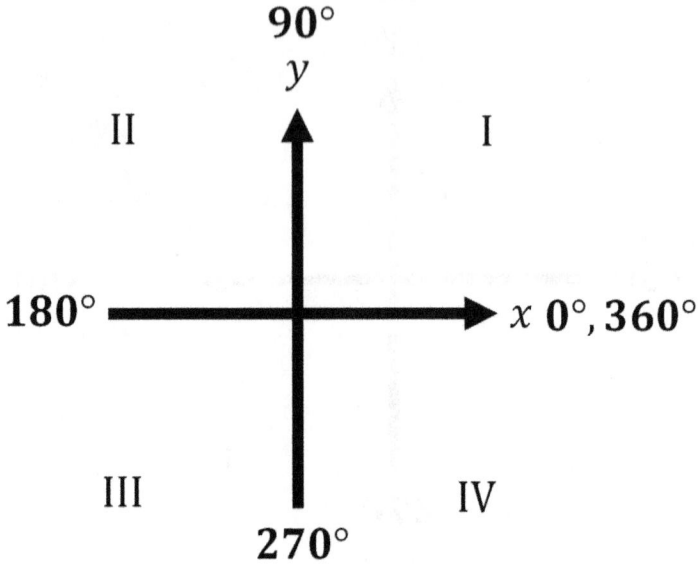

The four **Quadrants** correspond to the following ranges of angles:

∞ Quadrant I: $0° < \theta < 90°$
∞ Quadrant II: $90° < \theta < 180°$
∞ Quadrant III: $180° < \theta < 270°$
∞ Quadrant IV: $270° < \theta < 360°$

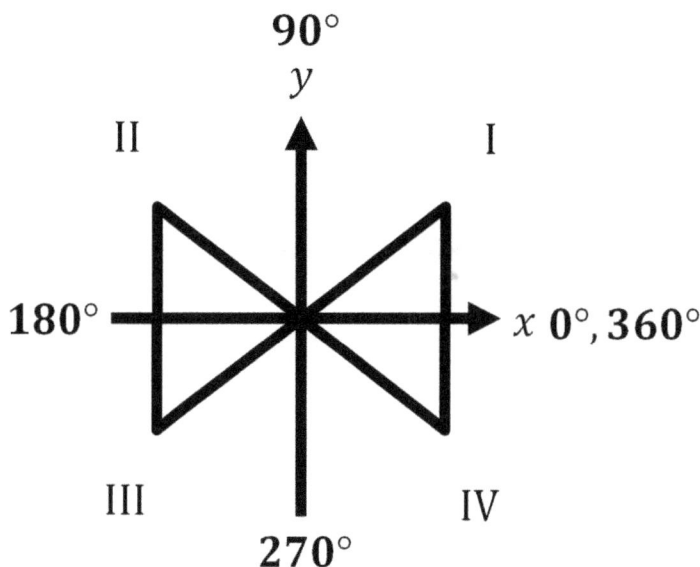

To relate the given angle to the reference angle, draw a triangle with its **base lying on the x-axis** in the appropriate Quadrant. The triangle will look like one of the 4 triangles shown above, depending on the Quadrant. This is illustrated with the examples that follow.

90°

y

II

I

180°

x 0°, 360°

III

IV

270°

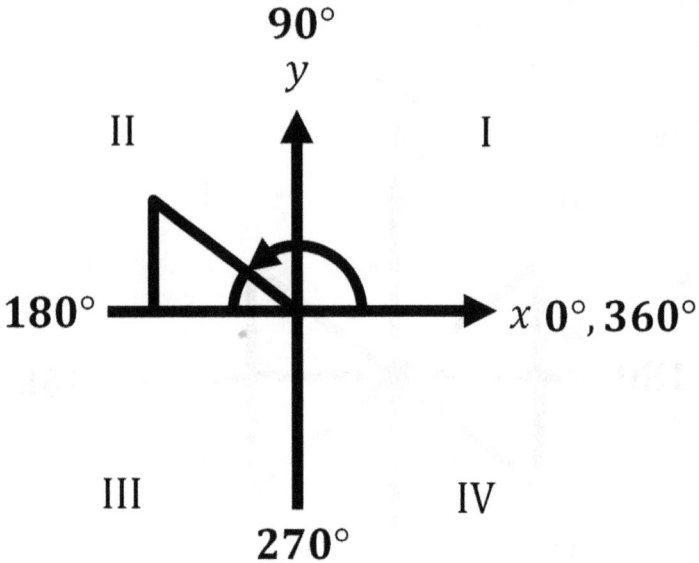

The **reference angle** (smaller angle shown above) is the smallest angle between the hypotenuse and the *x*-axis (or the −*x*-axis, whichever is smaller).

The **actual angle** (larger angle shown above) is counterclockwise from +*x*. In Quadrants II, III, and IV, the actual angle is greater than the reference angle.

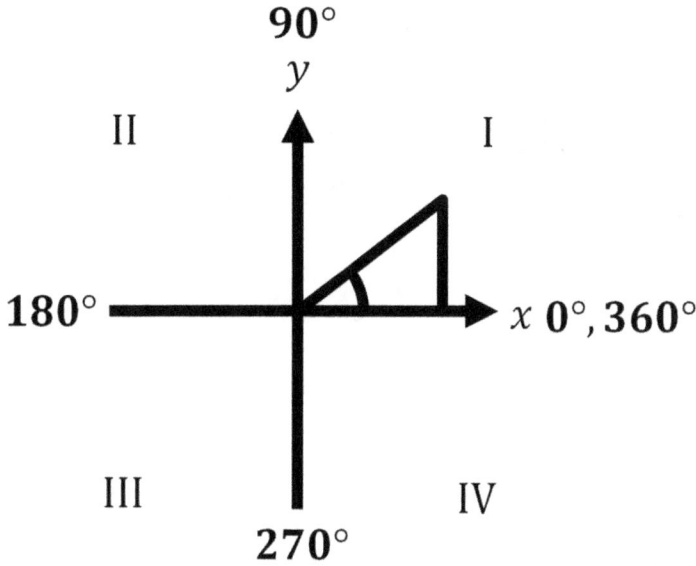

In Quadrant I, there is no distinction between the angle and the reference angle; they are the same in Quadrant I.

$$\theta_I = \theta_{ref}$$

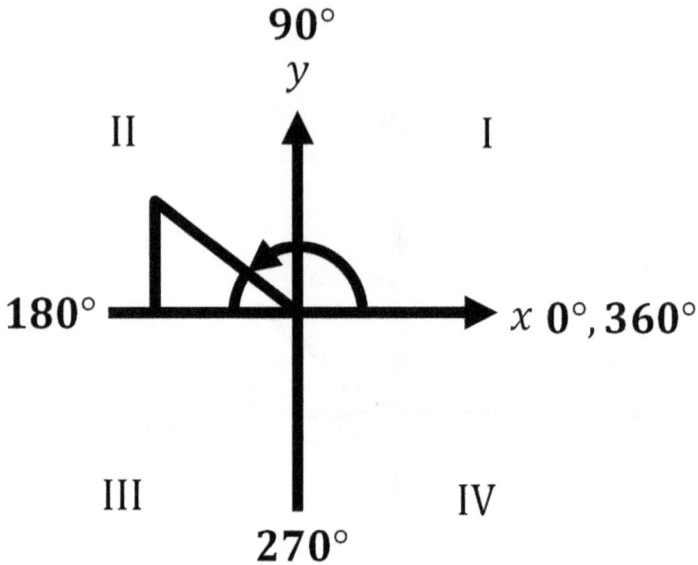

In Quadrant II, the angle and reference angle are supplementary. The angle (larger angle shown above) is measured counterclockwise from $+x$. The smaller angle shown above is the reference angle.

$$\theta_{II} = 180° - \theta_{ref}$$
$$\theta_{ref} = 180° - \theta_{II}$$

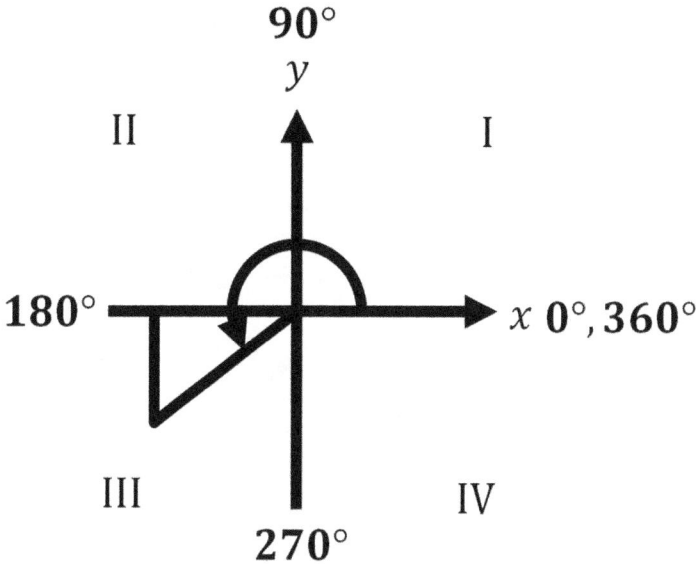

In Quadrant III, the angle is 180° more than the reference angle. The angle (larger angle shown above) is measured counterclockwise from $+x$.

$$\theta_{III} = 180° + \theta_{ref}$$
$$\theta_{ref} = \theta_{III} - 180°$$

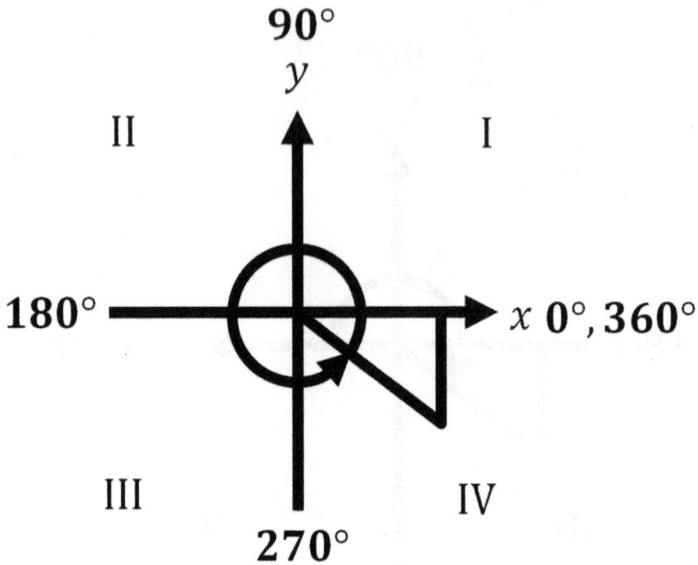

In Quadrant IV, the angle is 360° minus than the reference angle. The angle (larger angle shown above) is measured counterclockwise from $+x$. The smaller angle shown above is the reference angle.

$$\theta_{IV} = 360° - \theta_{ref}$$
$$\theta_{ref} = 360° - \theta_{IV}$$

To **find the actual angle**, given the reference angle, use these formulas:

$$\theta_I = \theta_{ref}$$
$$\theta_{II} = 180° - \theta_{ref}$$
$$\theta_{III} = 180° + \theta_{ref}$$
$$\theta_{IV} = 360° - \theta_{ref}$$

To **find the reference angle**, given the actual angle, use these formulas:

$$\theta_{ref} = \theta_I$$
$$\theta_{ref} = 180° - \theta_{II}$$
$$\theta_{ref} = \theta_{III} - 180°$$
$$\theta_{ref} = 360° - \theta_{IV}$$

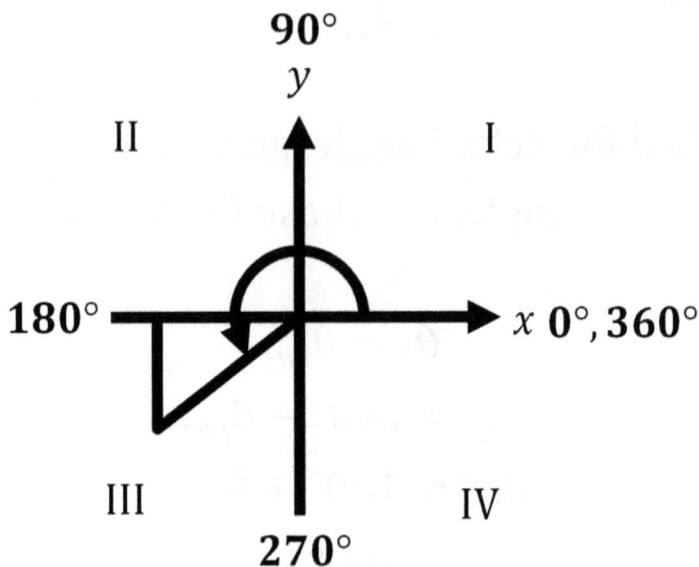

Example: What is the reference angle corresponding to 230°?

This angle lies in Quadrant III (because $180° < \theta < 270°$):

$$\theta_{ref} = \theta_{III} - 180°$$
$$\theta_{ref} = 230° - 180°$$
$$\theta_{ref} = 50°$$

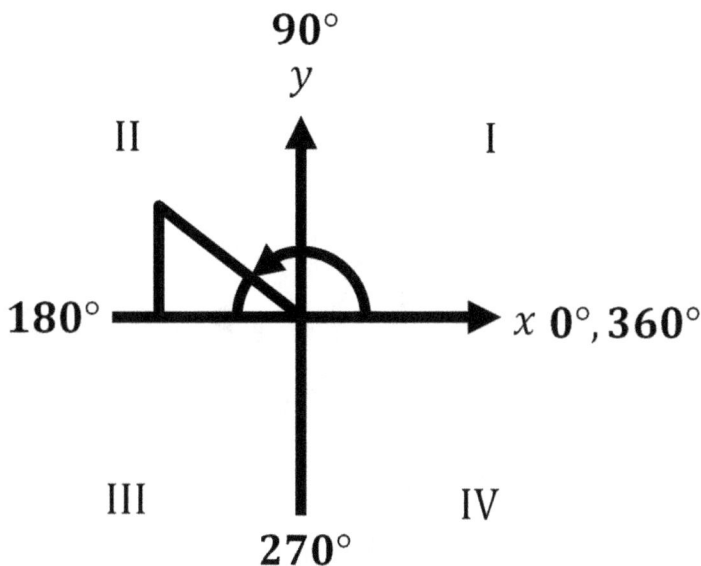

Example: What is the reference angle corresponding to 105°?

This angle lies in Quadrant II (because $90° < \theta < 180°$):

$$\theta_{ref} = 180° - \theta_{II}$$
$$\theta_{ref} = 180° - 105°$$
$$\theta_{ref} = 75°$$

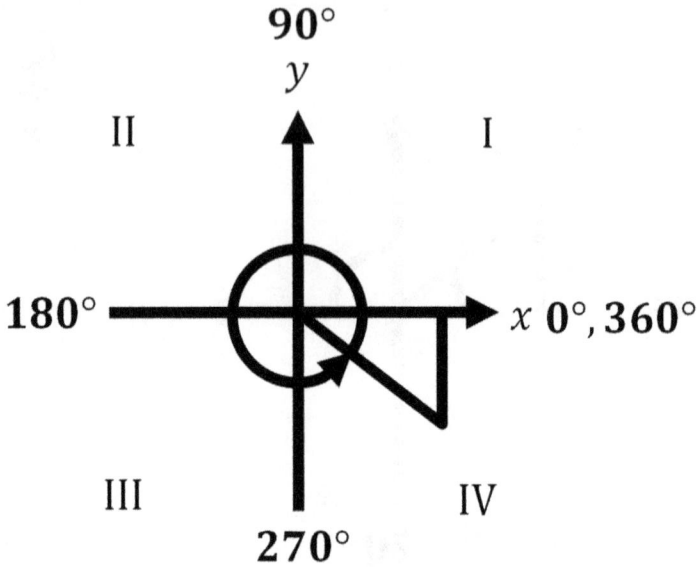

Example: What is the reference angle corresponding to 320°?

This angle lies in Quadrant IV (because $270° < \theta < 360°$):

$$\theta_{ref} = 360° - \theta_{IV}$$
$$\theta_{ref} = 360° - 320°$$
$$\theta_{ref} = 40°$$

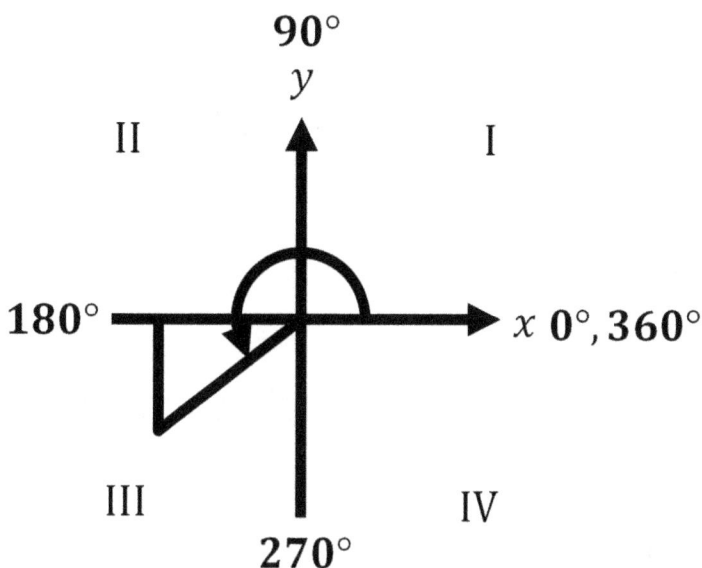

Example: What Quadrant III angle has a reference angle of 35°?

Notice how this example differs from the previous examples. This time, use a formula with θ_{ref} on the right side:

$$\theta_{III} = 180° + \theta_{ref}$$
$$\theta_{ref} = 180° + 35°$$
$$\theta_{ref} = 215°$$

Practice

Check your answers on page 132.

1. Find the reference angle corresponding to the angle given.

(A) 140°   (B) 310°   (C) 115°

(D) 190°   (E) 80°   (F) 290°

2. Given the reference angle, find the angle in the specified Quadrant.

(A) 55°, QII   (B) 70° QIII

(C) 40°, QIV   (D) 15°, QI

(E) 5°, QII   (F) 25°, QIII

5 THE UNIT CIRCLE

This chapter introduces the concept of the unit circle. Topics include:

∞ what the unit circle is
∞ (x, y) coordinates
∞ how to use the unit circle
∞ trig functions for the unit circle

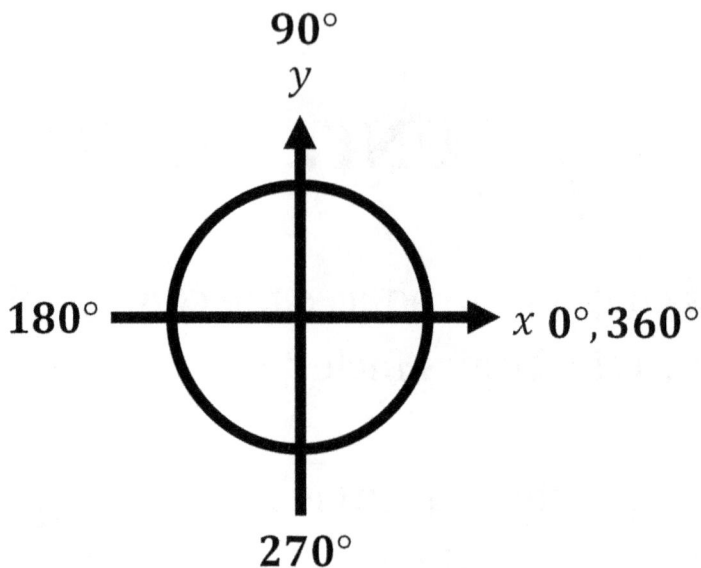

The **unit circle** is a circle with a radius of 1 unit lying in the xy plane, with its center at the origin.

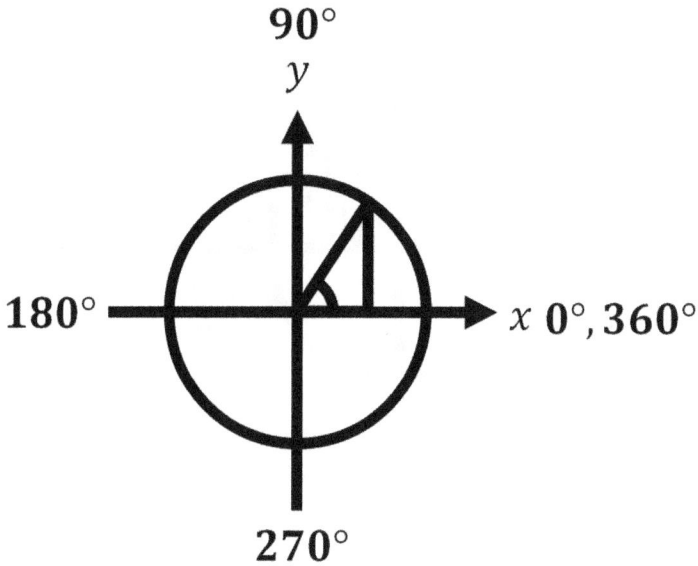

The unit circle makes the trig simpler when the hypotenuse of a right triangle is a radius of the unit circle. In this case, the **hypotenuse equals 1 unit.**

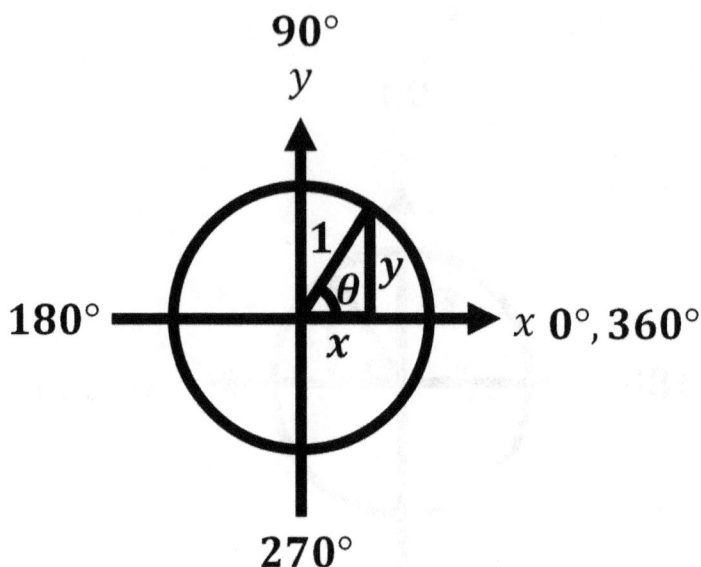

For a triangle drawn on unit circle as shown above:

∞ the hypotenuse equals 1
∞ the adjacent equals x
∞ the opposite equals y.

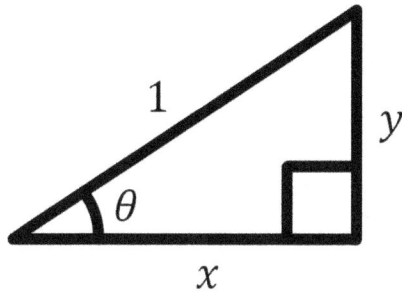

Since the hypotenuse is 1 on the unit circle, the basic trig functions simplify to:

$$\sin\theta = \frac{opp}{hyp} = \frac{y}{1} = y$$

$$\cos\theta = \frac{adj}{hyp} = \frac{x}{1} = x$$

$$\tan\theta = \frac{opp}{adj} = \frac{y}{x}$$

For the unit circle:

∞ **sin** $\boldsymbol{\theta}$ equals y

∞ **cos** $\boldsymbol{\theta}$ equals x

∞ **tan** $\boldsymbol{\theta}$ equals $\dfrac{y}{x}$

You can ignore the hypotenuse when working with the unit circle, since the hypotenuse is 1.

The unit circle makes it convenient to figure out the signs (+ or −) of the trig functions in Quadrants II-IV: Simply look at the signs of x and y.

∞ for x, right is + and left is −

∞ for y, up is + and down is −

$$90°$$
$$y$$

Quad. II
y is $+$

Quad. I
y is $+$

$$180°$$

$$x\ 0°, 360°$$

Quad. III
y is $-$

Quad. IV
y is $-$

$$270°$$

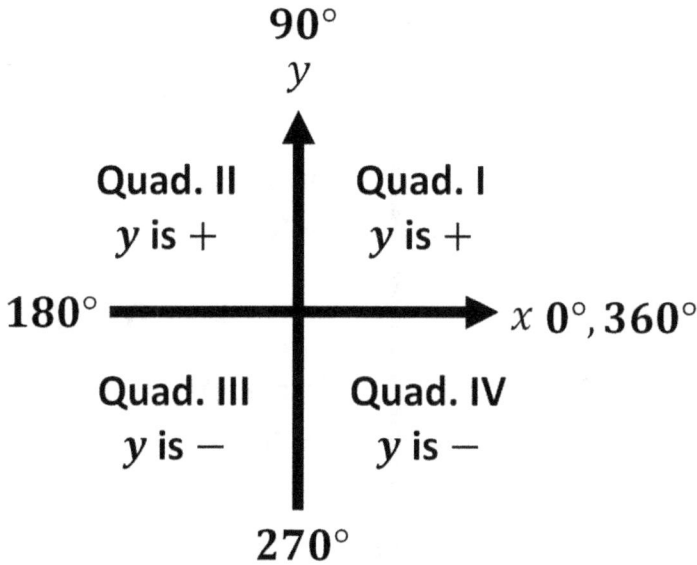

Since **sin θ** equals **y** on the unit circle, and since **y** is positive in Quadrants I/II and negative in Quadrants III/IV:

∞ **sin θ** is $+$ in Quadrants I/II
∞ **sin θ** is $-$ in Quadrants III/IV

90°

y

Quad. II
x is $-$

Quad. I
x is $+$

180°

x 0°, 360°

Quad. III
x is $-$

Quad. IV
x is $+$

270°

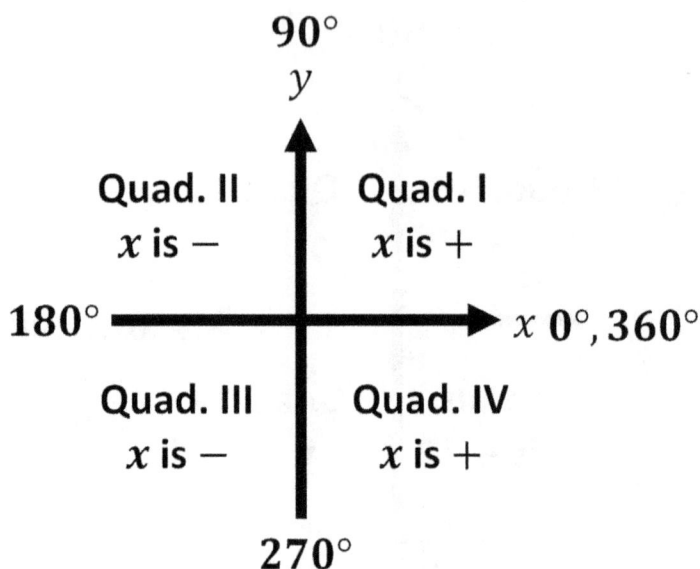

Since $\cos\theta$ equals x on the unit circle, and since x is positive in Quadrants I/IV and negative in Quadrants II/III:

∞ $\cos\theta$ is $+$ in Quadrants I/IV

∞ $\cos\theta$ is $-$ in Quadrants II/III

90°

y

Quad. II

x is −

y is +

Quad. I

x is +

y is +

180°

x 0°, 360°

Quad. III

x is −

y is −

Quad. IV

x is +

y is −

270°

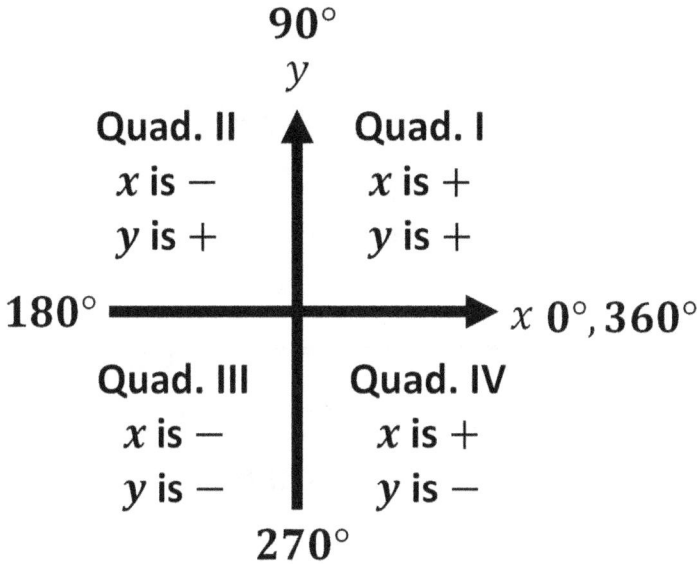

Since **tan θ** equals $\frac{y}{x}$ on the unit circle, tangent is positive when x and y are both positive or both negative, and tangent is negative otherwise:

∞ **tan θ** is + in Quadrants I/III

∞ **tan θ** is − in Quadrants II/IV

90°

Quad. II y **Quad. I**
sin θ is $+$ **sin θ is $+$**
cos θ is $-$ **cos θ is $+$**
tan θ is $-$ **tan θ is $+$**

180° ————————————➤ x **0°, 360°**

Quad. III   **Quad. IV**
sin θ is $-$ **sin θ is $-$**
cos θ is $-$ **cos θ is $+$**
tan θ is $+$ 270° tan θ is $-$

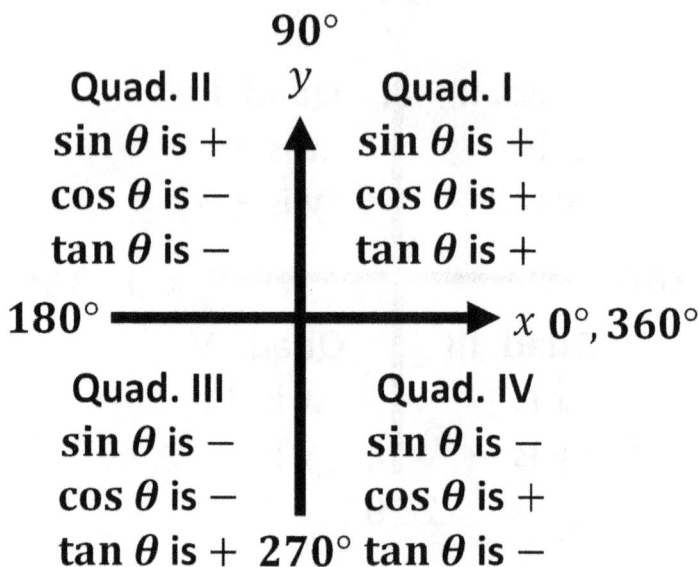

∞ **sin θ** is $+$ in Quadrants I/II
∞ **sin θ** is $-$ in Quadrants III/IV
∞ **cos θ** is $+$ in Quadrants I/IV
∞ **cos θ** is $-$ in Quadrants II/III
∞ **tan θ** is $+$ in Quadrants I/III
∞ **tan θ** is $-$ in Quadrants II/IV

90°

y

Quad. II
sin +

Quad. I
all +

180° ————————————→ *x* **0°, 360°**

Quad. III
tan +

Quad. IV
cos +

270°

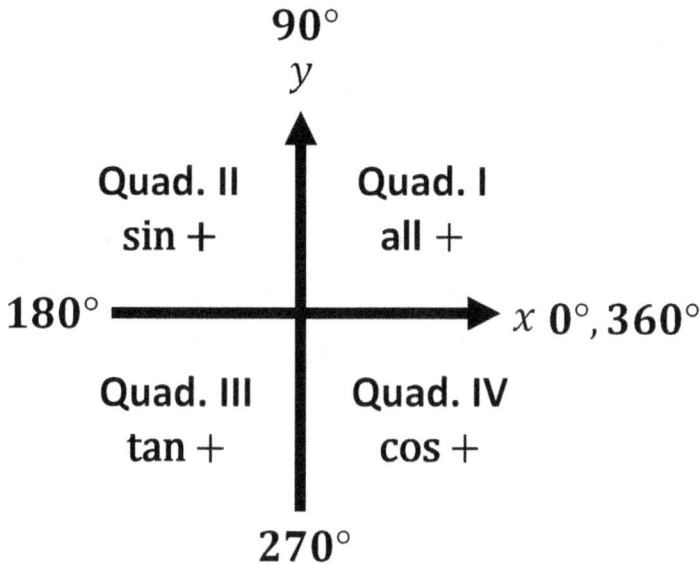

It is simpler to focus on which functions are positive in a given Quadrant:

∞ all are + in Quadrant I

∞ **sin** θ is + in Quadrant II

∞ **tan** θ is + in Quadrant III

∞ **cos** θ is + in Quadrant IV

Memory Tip: Remember the following silly sentence.

Apes Study Trig Calculations
∞ **all** are + in Quadrant I
∞ **sin** θ is + in Quadrant II
∞ **tan** θ is + in Quadrant III
∞ **cos** θ is + in Quadrant IV

The first letter of each word (**A**pes **S**tudy **T**rig **C**alculations) stands for **all**, **sin**, **tan**, or **cos**.

This can help you remember which function is positive in which Quadrant. The other functions are therefore negative in that Quadrant.

For example, since tan θ is + in Quadrant III, both sin θ and cos θ are negative in Quadrant III.

Practice

Check your answers on page 133.

Indicate whether each function is positive or negative for the given angle.

1. sin 123°   2. cos 228°   3. tan 346°

4. cos 96°   5. tan 12°   6. sin 199°

Indicate whether each function is positive or negative for the given angle.

7. tan 204°   8. sin 1°   9. cos 291°

10. sin 357°   11. tan 133°   12. cos 42°

6 QUADRANTS II-IV

This chapter focuses on evaluating the basic trig functions at angles in Quadrants II, III, and IV. Topics include:

- ∞ relating functions to Quadrant I
- ∞ how to combine sign with θ_{ref}
- ∞ a summary of equations for θ_{ref}*
- ∞ a summary of signs*
- ∞ examples in Quadrants II-IV

* See Chapter 4 to learn more about the reference angle. See Chapter 5 for to learn more about signs in Quadrants II-IV.

A trig function evaluated for an angle in Quadrants II thru IV can be related to a trig function evaluated for a Quadrant I angle using these steps:

∞ determine the sign (+ or −)
∞ determine the reference angle
∞ evaluate the trig function at θ_{ref}
∞ combine this with the sign

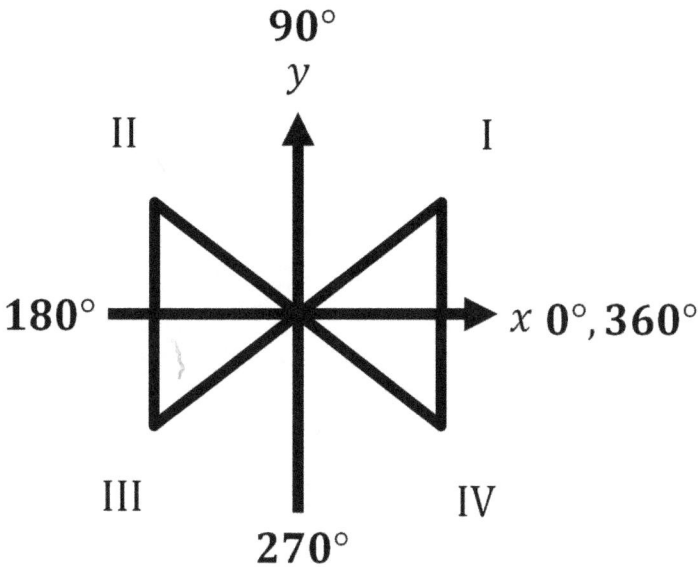

Following is a summary of the equations for finding the reference angle corresponding to a Quadrant II-IV angle. See Chapter 4 for more details.

$$\theta_{ref} = \theta_I$$
$$\theta_{ref} = 180° - \theta_{II}$$
$$\theta_{ref} = \theta_{III} - 180°$$
$$\theta_{ref} = 360° - \theta_{IV}$$

90°

y

Quad. II
sin +

Quad. I
all +

180° ————————————→ *x* **0°, 360°**

Quad. III
tan +

Quad. IV
cos +

270°

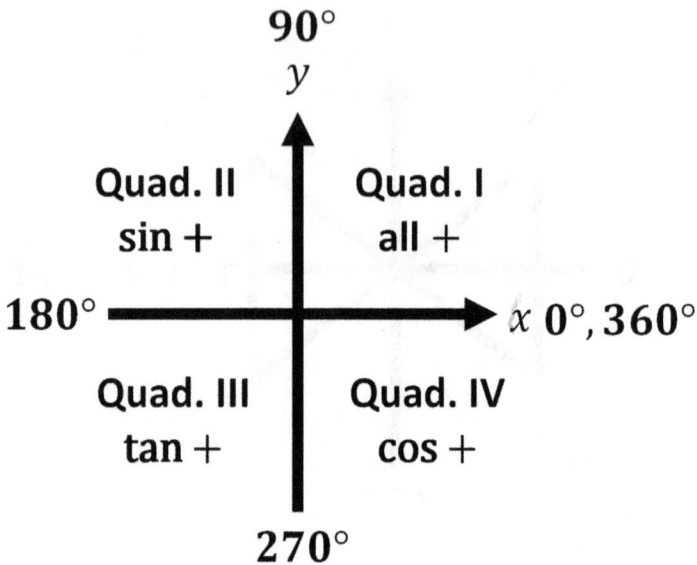

Following is a summary of which trig functions have which sign in which Quadrant. See Chapter 5 for more info.

∞ all are + in Quadrant I
∞ **sin θ** is + in Quadrant II
∞ **tan θ** is + in Quadrant III
∞ **cos θ** is + in Quadrant IV

Example: Calculate cos 120°.

120° lies in Quadrant II because it's between 90° and 180°.

Cosine is negative in Quadrant II, so cos 120° will be negative.

The reference angle is:

$$\theta_{ref} = 180° - \theta_{II} = 180° - 120° = 60°$$

Recall that $\cos 60° = \dfrac{1}{2}$.

Therefore,

$$\cos 120° = -\cos 60° = -\dfrac{1}{2}$$

Example: Calculate $\tan 210°$.

$210°$ lies in Quadrant III because it's between $180°$ and $270°$.

Tangent is positive in Quadrant III, so $\tan 210°$ will be positive.

The reference angle is:

$$\theta_{ref} = \theta_{III} - 180° = 210° - 180° = 30°$$

Recall that $\tan 30° = \frac{\sqrt{3}}{3}$.

Therefore,

$$\tan 210° = +\tan 30° = \frac{\sqrt{3}}{3}$$

Example: Calculate $\sin 315°$.

$315°$ lies in Quadrant IV because it's between $270°$ and $360°$.

Sine is negative in Quadrant IV, so $\sin 315°$ will be negative.

The reference angle is:

$$\theta_{ref} = 360° - \theta_{IV} = 360° - 315° = 45°$$

Recall that $\sin 45° = \dfrac{\sqrt{2}}{2}$.

Therefore,

$$\sin 315° = -\sin 45° = -\dfrac{\sqrt{2}}{2}$$

Practice

Check your answers on page 134.

1. $\sin 210° = ?$ 2. $\cos 315° = ?$

3. $\tan 120° = ?$ 4. $\cos 150° = ?$

5. $\sin 300° = ?$ 6. $\tan 180° = ?$

7. $\cos 270° = ?$ 8. $\sin 120° = ?$

9. $\sin 330° = ?$

10. $\tan 330° = ?$

11. $\cos 135° = ?$

12. $\sin 270° = ?$

13. $\tan 240° = ?$ 14. $\cos 180° = ?$

15. $\sin 225° = ?$ 16. $\tan 315° = ?$

7 SECANT, COSECANT, AND COTANGENT

The secant, cosecant, and cotangent functions are related to the sine, cosine, and tangent. This chapter covers:

∞ definitions of $\sec\theta$, $\csc\theta$, and $\cot\theta$
∞ relating to $\sin\theta$, $\cos\theta$, and $\tan\theta$
∞ remembering the correspondence

The secant ($\sec \theta$), cosecant ($\csc \theta$), and cotangent ($\cot \theta$) functions are reciprocals of other trig functions:

$$\sec \theta = \frac{1}{\cos \theta}$$

$$\csc \theta = \frac{1}{\sin \theta}$$

$$\cot \theta = \frac{1}{\tan \theta}$$

Note that cosine does not correspond to cosecant; these co's do not go together.

Rather, secant is the reciprocal of cosine, while cosecant is the reciprocal of sine.

Remember that the co's (cosine and cosecant) do **not** go together.

Here is how secant, cosecant, and cotangent relate to right triangles:

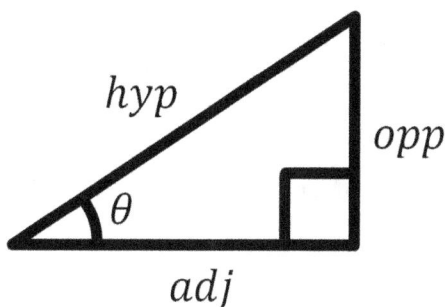

$$\textbf{sec } \boldsymbol{\theta} = \frac{hyp}{adj} = \frac{1}{\cos \theta}$$

$$\textbf{csc } \boldsymbol{\theta} = \frac{hyp}{opp} = \frac{1}{\sin \theta}$$

$$\textbf{cot } \boldsymbol{\theta} = \frac{adj}{opp} = \frac{1}{\tan \theta}$$

Finding secant, cosecant, and cotangent of an angle is easy: Just find the related trig function (cosine, sine, or tangent) and take its **reciprocal**.

$$\sec \theta = \frac{1}{\cos \theta}$$

$$\csc \theta = \frac{1}{\sin \theta}$$

$$\cot \theta = \frac{1}{\tan \theta}$$

Here is a brief review of **reciprocals**.

To find the reciprocal of a fraction, just invert the numerator and denominator.

For example, the reciprocal of 3/4 is 4/3. The old numerator (3) became the new denominator, while the old denominator (4) became the new numerator.

As another example, the reciprocal of 1/2 is 2/1, which reduces to 2.

To find the reciprocal of an integer, just write 1 divided by it.

For example, the reciprocal of 4 is 1/4 and the reciprocal of 7 is 1/7.

Example: Evaluate sec 60°.

$$\sec 60° = \frac{1}{\cos 60°} = \frac{1}{1/2} = 2$$

Recall the rule for dividing by a fraction: Multiply by its reciprocal:

$$\frac{1}{1/2} = 1 \times \frac{2}{1} = 1 \times 2 = 2$$

Example: Evaluate csc 45°.

$$\csc 45° = \frac{1}{\sin 45°} = \frac{1}{\frac{\sqrt{2}}{2}} = \frac{2}{\sqrt{2}}$$

$$= \frac{2}{\sqrt{2}}\frac{\sqrt{2}}{\sqrt{2}} = \frac{2\sqrt{2}}{2} = \sqrt{2}$$

Example: Evaluate $\cot 120°$.

$$\cot 120° = \frac{1}{\tan 120°} = -\frac{1}{\tan 60°}$$

$$= -\frac{1}{\sqrt{3}} = -\frac{1}{\sqrt{3}}\frac{\sqrt{3}}{\sqrt{3}} = -\frac{\sqrt{3}}{3}$$

Example: Evaluate $\sec 330°$.

$$\sec 330° = \frac{1}{\cos 330°} = +\frac{1}{\cos 30°}$$

$$= \frac{1}{\sqrt{3}/2} = \frac{2}{\sqrt{3}}\frac{\sqrt{3}}{\sqrt{3}} = \frac{2\sqrt{3}}{3}$$

These examples involve angles from Quadrants II-IV, which requires figuring out the sign and reference angle (as described in Chapter 6).

Practice

Check your answers on page 135.

1. csc $30° = $?   2. sec $30° = $?

3. cot $60° = $?   4. sec $120° = $?

5. $\csc 315° = ?$　　　　6. $\cot 180° = ?$

7. $\sec 225° = ?$　　　　8. $\csc 120° = ?$

9. $\cot 135° = ?$ 10. $\sec 300° = ?$

11. $\sec 0° = ?$ 12. $\cot 240° = ?$

13. $\csc 210° = ?$ 14. $\sec 45° = ?$

15. $\cot 30° = ?$ 16. $\csc 270° = ?$

8 INVERSE TRIG FUNCTIONS

This chapter introduces the inverse trig functions, including:

- ∞ what the inverse means
- ∞ how it differs from the reciprocal
- ∞ how to calculate the inverses
- ∞ two possible answers
- ∞ how to find both answers

The **inverse** basically means to do the opposite. A trig function takes an angle and returns a ratio. For example,

$$\sin 30° = \frac{1}{2}$$

Here, we take the sine of an angle, 30°, and get a fraction, $\frac{1}{2}$, as the answer.

The inverse sine function does the opposite. The inverse sine function takes a ratio and returns an angle. The opposite of the above equation is:

$$\sin^{-1}\left(\frac{1}{2}\right) = 30°$$

This **inverse sine** function asks, "What angle could you take the sine of and obtain $\frac{1}{2}$ as the answer?"

On a trig function, the superscript $^{-1}$ is not an exponent. The $^{-1}$ means to find the inverse; it does not mean to raise it to the power of -1.

In trig, the word **inverse** does not mean **reciprocal**:

∞ the **inverse** of $\sin \theta$ means $\sin^{-1}(x)$
∞ the **reciprocal** of $\sin \theta$ is $\csc \theta$
∞ $\sin^{-1}(x)$ is different from $\dfrac{1}{\sin x}$
∞ $\csc \theta$ is the same as $\dfrac{1}{\sin \theta}$
∞ in $\sin^{-1}(x)$, x is a ratio
∞ in $\sin \theta$, θ is an angle

In multiplication, the multiplicative inverse and reciprocal are identical: $x^{-1} = \dfrac{1}{x}$. However, this is not true in trig: $\sin^{-1}(x) \neq \dfrac{1}{\sin x}$.

What does inverse mean?

Here are some examples of what the inverse trig functions mean:

∞ $\sin^{-1}\left(\frac{\sqrt{2}}{2}\right)$ means, "What angle could you take the sine of and obtain $\frac{\sqrt{2}}{2}$ as the answer?"

∞ $\cos^{-1}\left(\frac{1}{2}\right)$ means, "What angle could you take the cosine of and obtain $\frac{1}{2}$ as the answer?"

∞ $\tan^{-1}(1)$ means, "What angle could you take the tangent of and obtain 1 as the answer?"

Note: There is a distinction between an angle (in degrees or radians) and a ratio (which has no units).

You can **take the sine of an angle**, like $\sin 60°$. You can't take the sine of a unitless ratio. If you write something like $\sin \frac{1}{2}$, you made a mistake (unless you mean half a degree or half a radian). You can take the sine of an angle in radians (we'll learn about radians in Chapter 9).

In contrast, you **take the inverse sign of a ratio**, like $\sin^{-1}\left(\frac{1}{2}\right)$. You can't take the inverse sine of an angle. If you write something like $\sin^{-1} 60°$, you made a mistake.

The equation

$$\sin 60° = \frac{\sqrt{3}}{2}$$

says that the sine of a 60° angle equals the ratio $\frac{\sqrt{3}}{2}$.

The equation

$$\sin^{-1}\left(\frac{\sqrt{3}}{2}\right) = 60°$$

asks what angle you could take the sine of in order to obtain the ratio $\frac{\sqrt{3}}{2}$. One answer equals 60° because

$$\sin 60° = \frac{\sqrt{3}}{2}$$

$$90°$$
$$y$$

Quad. II
sin +

Quad. I
all +

$$180°$$

$$x \; 0°, 360°$$

Quad. III
tan +

Quad. IV
cos +

$$270°$$

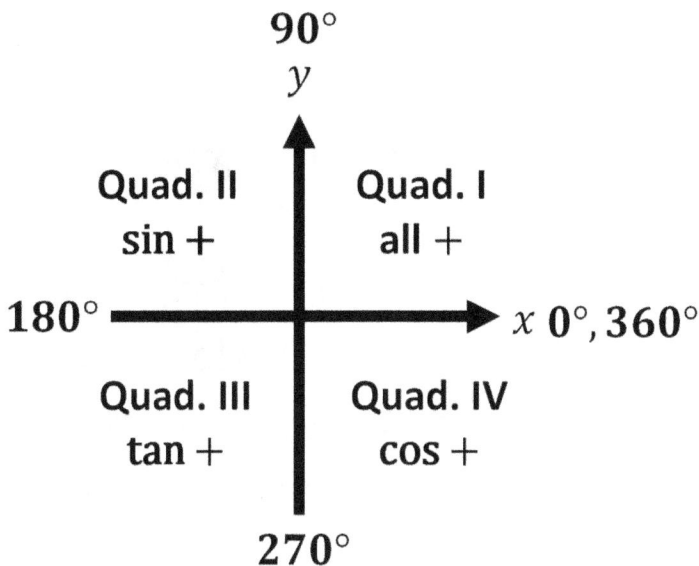

In general, an inverse trig function has not just one, but **two different answers**.

Why? Because each trig function is positive in two Quadrants, and is also negative in two Quadrants. So there are generally two Quadrants that will yield the correct answer.

This is shown in the following examples.

How to find $\sin^{-1} x$, $\cos^{-1} x$, and $\tan^{-1} x$:

∞ First, ignore the sign of x.

∞ Find the reference angle by asking, "What angle can you evaluate the trig function at and obtain $|x|$ as the answer?"

∞ Now look at the sign of x. In which two Quadrants does the given trig function have that sign ($+$ or $-$)?

∞ Use the formulas (see Chapter 6) to find the angles that correspond to the reference angle.

Study the following examples to see how this strategy is applied.

Example: Find $\sin^{-1}\left(\frac{1}{2}\right)$.

Step 1: The reference angle is 30° because $\sin 30° = \frac{1}{2}$.

Step 2: The argument of $\sin^{-1}$ is positive. The sine function is positive in Quadrants I and II.

Step 3: Use the reference angle to find the Quadrant I and II equivalents:

$$\theta_I = \theta_{ref} = 30°$$
$$\theta_{II} = 180° - \theta_{ref} = 180° - 30° = 150°$$

The two answers are 30° and 150°.

Example: Find $\cos^{-1}\left(-\frac{\sqrt{2}}{2}\right)$.

Step 1: The reference angle is 45° because $\cos 45° = \frac{\sqrt{2}}{2}$. (Ignore the − sign in Step 1.)

Step 2: The argument of $\cos^{-1}$ is negative. The cosine function is negative in Quadrants
II and III.

Step 3: Use the reference angle to find the Quadrant II and III equivalents:

$$\theta_{II} = 180° - \theta_{ref} = 180° - 45° = 135°$$
$$\theta_{III} = 180° + \theta_{ref} = 180° + 45° = 225°$$

The two answers are 135° and 225°.

Example: Find $\tan^{-1}(-1)$.

Step 1: The reference angle is 45° because $\tan 45° = 1$. (Ignore the − sign in Step 1.)

Step 2: The argument of $\tan^{-1}$ is negative. The tangent function is negative in Quadrants II and IV.

Step 3: Use the reference angle to find the Quadrant II and IV equivalents:

$$\theta_{II} = 180° - \theta_{ref} = 180° - 45° = 135°$$
$$\theta_{IV} = 360° - \theta_{ref} = 360° - 45° = 315°$$

The two answers are 135° and 315°.

How to find $\mathbf{sec^{-1}}\, \boldsymbol{x}$, $\mathbf{csc^{-1}}\, \boldsymbol{x}$, and $\mathbf{cot^{-1}}\, \boldsymbol{x}$:

First, find $\dfrac{1}{x}$.

For $\sec^{-1} x$, find $\cos^{-1}\left(\dfrac{1}{x}\right)$.

For $\csc^{-1} x$, find $\sin^{-1}\left(\dfrac{1}{x}\right)$.

For $\cot^{-1} x$, find $\tan^{-1}\left(\dfrac{1}{x}\right)$.

Now use the strategy to find the inverse sine, cosine, or tangent.

This method is illustrated in the examples that follow.

Example: Find $\sec^{-1}(-2)$.

Step 1: The reciprocal of -2 is $-\frac{1}{2}$. Thus,

$\sec^{-1}(-2) = \cos^{-1}(-\frac{1}{2})$.

Step 2: The reference angle is $60°$ because $\cos 60° = \frac{1}{2}$. (Ignore the $-$ sign in Step 2.)

Step 3: The argument of $\cos^{-1}$ is negative. The cosine function is negative in Quadrants II and III.

Step 4: Use the reference angle to find the Quadrant II and III equivalents:

$$\theta_{II} = 180° - \theta_{ref} = 180° - 60° = 120°$$
$$\theta_{III} = 180° + \theta_{ref} = 180° + 60° = 240°$$

The two answers are $120°$ and $240°$.

Practice

Check your answers on page 136.

1. $\sin^{-1}\left(\frac{\sqrt{3}}{2}\right) = ?$ 2. $\cos^{-1}\left(\frac{1}{2}\right) = ?$

3. $\sec^{-1}\left(\sqrt{2}\right) = ?$ 4. $\tan^{-1}\left(\frac{\sqrt{3}}{3}\right) = ?$

5. $\csc^{-1}(2) = ?$

6. $\sin^{-1}\left(-\frac{\sqrt{2}}{2}\right) = ?$

7. $\cot^{-1}\left(-\sqrt{3}\right) = ?$

8. $\sec^{-1}\left(-\frac{2\sqrt{3}}{3}\right) = ?$

9. $\cos^{-1}\left(-\frac{\sqrt{3}}{2}\right) = ?$ 10. $\tan^{-1}(0) = ?$

11. $\sin^{-1}(-1) = ?$ 12. $\cos^{-1}\left(\frac{\sqrt{2}}{2}\right) = ?$

13. $\cot^{-1}(1) = ?$

14. $\csc^{-1}\left(-\sqrt{2}\right) = ?$

15. $\sec^{-1}(-1) = ?$

16. $\tan^{-1}\left(-\sqrt{3}\right) = ?$

9 RADIANS

Angles are commonly expressed in degrees or radians. The topics of this chapter include:

∞ what a radian is
∞ converting radians to degrees
∞ converting degrees to radians
∞ a chart with special angles

Angles are often expressed in either degrees or radians.

Degrees are common when measuring angles with protractors. A degree is 1/360 of a complete circle.

Radians are common when working out calculations. The definition of the radian is related to the formula for arc length.

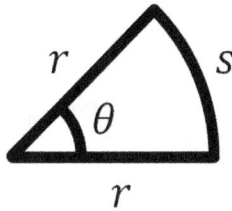

The equation for the **arc length** along a circle is

$$s = r\theta$$

where s represents arc length, r is the radius of a circle, and θ is the corresponding angle measured from the center of the circle, as shown above.

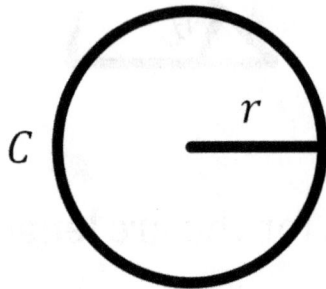

The equation for **circumference** is

$$C = 2\pi r$$

where C is the circumference (the complete distance around a circle).

The equation for arc length, $s = r\theta$, must agree with the equation for circumference, $C = 2\pi r$, when the arc length, s, happens to equal the circumference.

That is, if you walk around a circle exactly one time, your arc length equals your circumference.

This would be true if θ equals 2π for one complete revolution.

That is exactly how a radian is defined. **A radian is defined so that one complete revolution equates to 2π radians:**

$$1 \text{ revolution} = 2\pi \text{ radians}$$

One revolution corresponds to 360 degrees and also corresponds to 2π radians.

Thus, the conversion factor between degrees and radians is

$$360° = 2\pi \text{ rad}$$

Dividing both sides by 2, this **conversion factor** simplifies to

$$180° = \pi \text{ rad}$$

Here, rad is the abbreviation for radians and ° is the symbol for degrees.

To convert from **degrees to radians**, multiply by π rad and divide by 180°:

$$\frac{\pi \text{ rad}}{180°}$$

To convert from **radians to degrees**, multiply by 180° and divide by π rad:

$$\frac{180°}{\pi \text{ rad}}$$

These strategies are illustrated with the following examples.

Example: Convert 30° to radians.

$$30° = 30° \frac{\pi \text{ rad}}{180°} = \frac{\pi}{6} \text{ rad}$$

Example: Convert 120° to radians.

$$120° = 120° \frac{\pi \text{ rad}}{180°} = \frac{2\pi}{3} \text{ rad}$$

Example: Convert 360° to radians.

$$360° = 360° \frac{\pi \text{ rad}}{180°} = 2\pi \text{ rad}$$

Example: Convert $\dfrac{\pi}{2}$ rad to degrees.

$$\frac{\pi}{2}\ \text{rad} = \frac{\pi}{2}\ \text{rad}\,\frac{180°}{\pi\ \text{rad}} = 90°$$

Example: Convert $\dfrac{3\pi}{4}$ rad to degrees.

$$\frac{3\pi}{4}\ \text{rad} = \frac{3\pi}{4}\ \text{rad}\,\frac{180°}{\pi\ \text{rad}} = 135°$$

Example: Convert $\dfrac{4\pi}{3}$ rad to degrees.

$$\frac{4\pi}{3}\ \text{rad} = \frac{4\pi}{3}\ \text{rad}\,\frac{180°}{\pi\ \text{rad}} = 240°$$

Practice

Check your answers on page 137.

1. Convert from degrees to radians.

(A) 45°   (B) 60°   (C) 105°

(D) 225°   (E) 15°   (F) 270°

(G) 24°   (H) 210°   (I) 720°

2. Convert from radians to degrees.

(A) $\frac{\pi}{2}$ rad   (B) $\frac{\pi}{4}$ rad   (C) $\frac{5\pi}{6}$ rad

(D) $\frac{\pi}{9}$ rad   (E) $\frac{3\pi}{2}$ rad   (F) $\frac{\pi}{15}$ rad

(G) 3π rad   (H) $\frac{7\pi}{4}$ rad   (I) $\frac{5\pi}{3}$ rad

Related books by the author.

Want to learn more trig?

The learning continues with:

Learn or Review Trigonometry
Graphs, Trig Identities, Equations & More

Need more practice?

Develop fluency with:

Trigonometry Essentials
Practice Workbook with Answers

Memorize basic trig functions of special angles in Quadrants I-IV using:

Trigonometry Flash Cards

ANSWER KEY

Chapter 1 Answers

1. The angles of a triangle add up to 180°.

$$\beta = 55°, \theta = 65°$$

2. Use the Pythagorean Theorem.

$$c = 15, a = 3$$

Chapter 2 Answers

Use the Pythagorean Theorem to find the unknown side, then apply the definitions of the trig functions.

1. left: $\sin \theta = \dfrac{4}{5}$, $\cos \theta = \dfrac{3}{5}$, $\tan \theta = \dfrac{4}{3}$

right: $\sin \theta = \dfrac{\sqrt{2}}{2}$, $\cos \theta = \dfrac{\sqrt{2}}{2}$, $\tan \theta = 1$

2. left: $\sin \theta = \dfrac{\sqrt{5}}{5}$, $\cos \theta = \dfrac{2\sqrt{5}}{5}$, $\tan \theta = \dfrac{1}{2}$

right: $\sin \theta = \dfrac{1}{2}$, $\cos \theta = \dfrac{\sqrt{3}}{2}$, $\tan \theta = \dfrac{\sqrt{3}}{3}$

Chapter 3 Answers

1. $\sin 60° = \dfrac{\sqrt{3}}{2}$

2. $\cos 45° = \dfrac{\sqrt{2}}{2}$

3. $\tan 45° = 1$

4. $\sin 0° = 0$

5. $\cos 30° = \dfrac{\sqrt{3}}{2}$

6. $\tan 30° = \dfrac{\sqrt{3}}{3}$

7. $\cos 90° = 0$

8. $\sin 30° = \dfrac{1}{2}$

9. $\tan 60° = \sqrt{3}$

10. $\cos 60° = \dfrac{1}{2}$

11. $\sin 45° = \dfrac{\sqrt{2}}{2}$

12. $\tan 0° = 0$

13. $\cos 0° = 1$

14. $\tan 90° = \text{und.}$

15. $\sin 90° = 1$

16. $\sin 60° = \dfrac{\sqrt{3}}{2}$

Chapter 4 Answers

Either use the formulas for the reference angles, or draw the angle and work out the reference angle geometrically.

1. (A) 40°   (B) 50°   (C) 65°

(D) 10°   (E) 80°   (F) 70°

2. (A) 125°   (B) 250°

(C) 320°   (D) 15°

(E) 175°   (F) 205°

Chapter 5 Answers

Apply the rules for the signs of the trig functions in each Quadrant.

1. +   2. −   3. −

4. −   5. +   6. −

7. +   8. +   9. +

10. −   11. −   12. +

Chapter 6 Answers

1. $\sin 210° = -\dfrac{1}{2}$

2. $\cos 315° = \dfrac{\sqrt{2}}{2}$

3. $\tan 120° = -\sqrt{3}$

4. $\cos 150° = -\dfrac{\sqrt{3}}{2}$

5. $\sin 300° = -\dfrac{\sqrt{3}}{2}$

6. $\tan 180° = 0$

7. $\cos 270° = 0$

8. $\sin 120° = \dfrac{\sqrt{3}}{2}$

9. $\sin 330° = -\dfrac{1}{2}$

10. $\tan 330° = -\dfrac{\sqrt{3}}{3}$

11. $\cos 135° = -\dfrac{\sqrt{2}}{2}$

12. $\sin 270° = -1$

13. $\tan 240° = \sqrt{3}$

14. $\cos 180° = -1$

15. $\sin 225° = -\dfrac{\sqrt{2}}{2}$

16. $\tan 315° = -1$

Chapter 7 Answers

1. $\csc 30° = 2$

2. $\sec 30° = \dfrac{2\sqrt{3}}{3}$

3. $\cot 60° = \dfrac{\sqrt{3}}{3}$

4. $\sec 120° = -2$

5. $\csc 315° = -\sqrt{2}$

6. $\cot 180° = $ und.

7. $\sec 225° = -\sqrt{2}$

8. $\csc 120° = \dfrac{2\sqrt{3}}{3}$

9. $\cot 135° = -1$

10. $\sec 300° = 2$

11. $\sec 0° = 1$

12. $\cot 240° = \dfrac{\sqrt{3}}{3}$

13. $\csc 210° = -2$

14. $\sec 45° = \sqrt{2}$

15. $\cot 30° = \sqrt{3}$

16. $\csc 270° = -1$

Chapter 8 Answers

1. 60°, 120°

2. 60°, 300°

3. 45°, 315°

4. 30°, 210°

5. 30°, 150°

6. 225°, 315°

7. 150°, 330°

8. 150°, 210°

9. 150°, 210°

10. 0°, 180°

11. 270°

12. 45°, 315°

13. 45°, 225°

14. 225°, 315°

15. 180°

16. 120°, 300°

Chapter 9 Answers

1. Multiply by $\frac{\pi \text{ rad}}{180°}$.

(A) $\frac{\pi}{4}$ rad   (B) $\frac{\pi}{3}$ rad   (C) $\frac{7\pi}{12}$ rad

(D) $\frac{5\pi}{4}$ rad   (E) $\frac{\pi}{12}$ rad   (F) $\frac{3\pi}{2}$ rad

(G) $\frac{2\pi}{15}$ rad   (H) $\frac{7\pi}{6}$ rad   (I) 4π rad

2. Multiply by $\frac{180°}{\pi \text{ rad}}$.

(A) 90°   (B) 45°   (C) 150°

(D) 20°   (E) 270°   (F) 12°

(G) 540°   (H) 315°   (I) 300°

ABOUT THE AUTHOR

Chris McMullen is a physics instructor at Northwestern State University of Louisiana and also an author of academic books. Whether in the classroom or as a writer, Dr. McMullen loves sharing knowledge and the art of motivating and engaging students.

He earned his Ph.D. in phenomenological high-energy physics (particle physics) from Oklahoma State University in 2002. Originally from California, Dr. McMullen earned his Master's degree from California State University, Northridge, where his thesis was in the field of electron spin resonance.

As a physics teacher, Dr. McMullen observed that many students lack fluency in fundamental math skills. In an effort to help students of all ages and levels master basic math skills, he published a series of math workbooks on arithmetic, fractions, and algebra called the Improve Your Math Fluency Series. Dr. McMullen has also published a variety of science books, including introductions to basic astronomy and chemistry concepts in addition to physics textbooks.

Dr. McMullen is very passionate about teaching. Many students and observers have been impressed with the transformation that occurs when he walks into the classroom, and the interactive engaged discussions that he leads during class time. Dr. McMullen is well-known for drawing monkeys and using them in his physics examples and problems, using his creativity to inspire students. A stressed out student is likely to be told to throw some bananas at monkeys, smile, and think happy physics thoughts.

Author, Chris McMullen, Ph.D.

Improve Your Math Fluency

This series of math workbooks is geared toward practicing essential math skills:

- ∞ Algebra and trigonometry
- ∞ Fractions, decimals, and percents
- ∞ Long division
- ∞ Multiplication and division
- ∞ Addition and subtraction

TRIGONOMETRY

Essentials Practice Workbook with Answers

Master Basic Trig Skills

$$\cos(\alpha + \beta) =$$
$$\cos \alpha \cos \beta - \sin \alpha \sin \beta$$

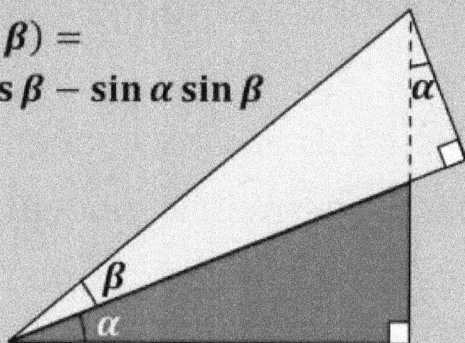

α

β

α

Improve Your Math Fluency Series

Chris McMullen, Ph.D.

Dr. McMullen has published a variety of **science** books, including:

∞ Basic astronomy concepts
∞ Basic chemistry concepts
∞ Creative physics problems
∞ Calculus-based physics

UNDERSTAND

BASIC

CHEMISTRY

CONCEPTS

(Large Size and Large Print Edition)

Chris McMullen, Ph.D.
Northwestern State University of Louisiana

YOU CaN

Chris McMullen enjoys solving puzzles. His favorite puzzle is Kakuro (kind of like a cross between crossword puzzles and Sudoku). He once taught a three-week summer course on puzzles.

If you enjoy mathematical pattern puzzles, you might appreciate:

300+ Mathematical Pattern Puzzles
Number Pattern Recognition & Reasoning

∞ pattern recognition
∞ visual discrimination
∞ analytical skills
∞ logic and reasoning
∞ analogies
∞ mathematics

300+

MATHEMATICAL
PATTERN
PUZZLES

NUMBER PATTERN RECOGNITION AND REASONING

CHRIS MCMULLEN, PH.D.

Chris McMullen has coauthored several word scramble books. This includes a cool idea called **VErBAl ReAcTiONS**. A VErBAl ReAcTiON expresses word scrambles so that they look like chemical reactions. Here is an example:

$$2\,C + U + 2\,S + Es \rightarrow S\,U\,C\,C\,Es\,S$$

The left side of the reaction indicates that the answer has 2 C's, 1 U, 2 S's, and 1 Es. Rearrange CCUSSEs to form SUCCEsS.

V Er B Al
Re Ac Ti O N S

Vanadium · Erbium · Boron · Aluminum
Rhenium · Actinium · Titanium · Oxygen · Nitrogen · Sulfur

Word Scrambles with a Chemical Flavor
MEDIUM
Rearrange Symbols from Chemistry's Periodic Table to Unscramble the Words

$$S + Ni + Ge + U \rightarrow \underline{Ge}\,\underline{Ni}\,\underline{U}\,\underline{S}$$

$$2C + N + 2I + P \rightarrow \underline{P}\,\underline{I}\,\underline{C}\,\underline{N}\,\underline{I}\,\underline{C}$$

$$Ti + C + Cr + P + Y \rightarrow \underline{Cr}\,Y\,\underline{P}\,\underline{Ti}\,\underline{C}$$

$$2C + U + 2S + Es \rightarrow \underline{S}\,\underline{U}\,\underline{C}\,\underline{C}\,\underline{Es}\,\underline{S}$$

Chris McMullen and Carolyn Kivett

FUN FOR WORD PUZZLE FANS! NO SCIENCE NEEDED!

Each answer to a **VErBAl ReAcTiON** is not merely a word, it's a chemical word. A chemical word is made up not of letters, but of elements of the periodic table. In this case, SUCCEsS is made up of sulfur (S), uranium (U), carbon (C), and Einsteinium (Es).

Another example of a chemical word is GeNiUS. It's made up of germanium (Ge), nickel (Ni), uranium (U), and sulfur (S).

If you enjoy anagrams and like science or math, these puzzles are tailor-made for you.